How to Build and Maintain a Drone from Scratch

The simplified method of constructing, handling, and maintaining a drone from scratch even as a as a novice or Professionals

BY,

Engr. Kris Kay

copyright@2021

TABLE OF CONTENT

CHAPTER ONE

INTRODUCTION TO BUILDING A DROME

Automated aeronautical vehicles have been around for quite a long time, yet they have accomplished the best fame lately with little business drones. The new alleged first individual view innovation gave us a remarkable encounter of flying and the headway of GPS frameworks in robots and it create an eye exposure to energetic people.

Obviously, drones are by all account not the only RC flying gadgets available, yet their dexterous multi-rotors and their ability to take astonishing photographs and record dazzling recordings during flight made them the most well known. That is the reason business drones are of higher interest nowadays by many people. Take your time and see how you can construct a DIY drone without any preparation?

In the world today, robots have wider scopes which vary in size, plan, and properties.

Visit popular online stores where the robots are being sold and locate some prepared to utilize a model that best suits you, considering the cost.

Great men will prefer purchasing a robot rather than building one. People who want to build their own drone without preparation will appreciate it.

The genuine test here is to locate the fundamental pieces and envision the robot structure yourself. This can be an amazingly muddled task, contingent upon the kind of robot you need to fabricate, and the materials essential. This book will take you through an overall layout of what it resembles to make a DIY quadcopter drone. This project is worth embarking on because the end fulfillment is more than justified, despite all the trouble!

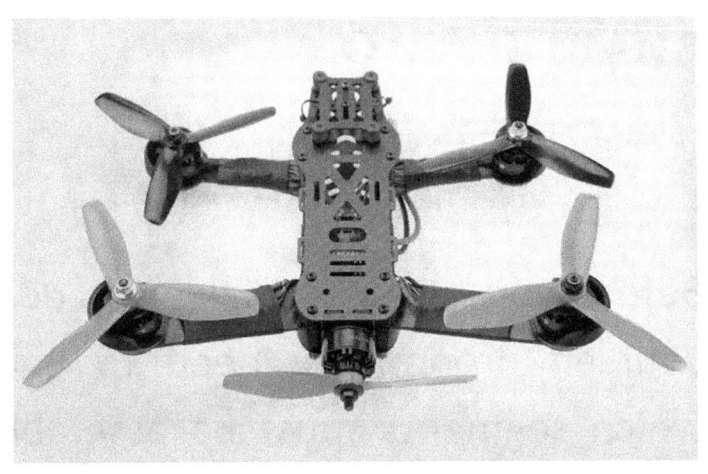

CHAPTER TWO

ESSENTIAL PARTS IN BUILDING A DRONE

Before you begin to make this do-it-yourself drone, you need to know which segments must be assembled to enable it fly easily.

Here are the essential parts needed to easily construct your own robot:

• **Frame**: there are two options

When it comes to getting a casing for your robot. You can decide to make it yourself or get it in an online shop. Getting the frame yourself is not too difficult as other persons may think but you will need resourceful materials like this book to undertake this step. It is advisable for you to use less weighty metal, plastics or wooden support. On the off chance that you pick a wooden edge, you'll need a wood board which is about 2.5 cm thick.

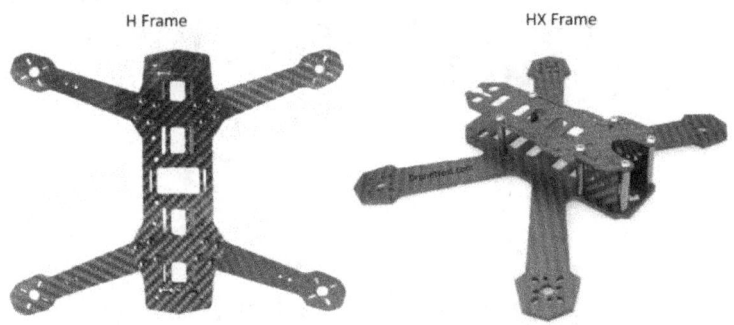

H Frame   HX Frame

• **Motors**: if you are building for a normal quad, 4 engines will be required altogether, while an octocopter needs eight engines to fly. The suggestion now is to use brushless engines since they are lighter on the battery.

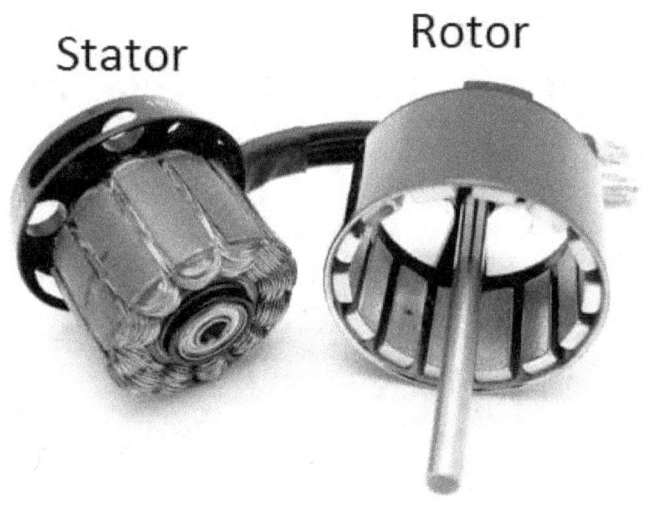

Stator   Rotor

If you are an architect who knows better functions of engine can purchase one from a store.

- **ESCs or electronic speed control**: these are refers to as fundamental bits of your robot as they function in conveying capacity to the engines. Electronic speed control is determine by the quantity of arms your robot will have.

- **Propellers**: When searching for the propellers, go for the one that can coordinate the edge of your robot. Wooden propellers cannot be found therefore make sure you pick a solid match for your project.

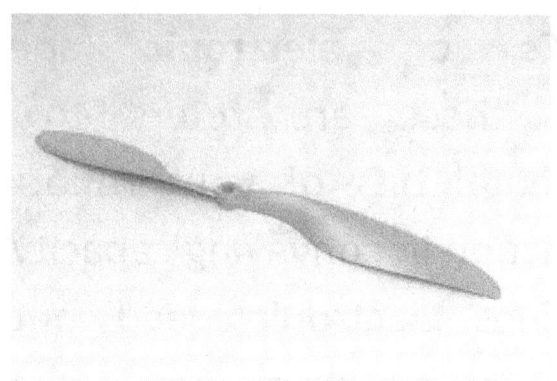

- **Connectors**: 3.5 mm connectors are required to weld the engines and ESCs. 4.5 mm connector is used for the force circulation board.

- **The power dispersion board** the speed control from the board to the battery is enhanced by the power dispersion board.

- **Batteries**: When purchasing batteries for your robot, you need to think about the battery lifespan. The most utilized batteries for this object are a Li-Po battery which has several ranges of force and their force varies.

• **Battery screen**: This is not seen as a rudimentary thing but the screen is very valuable since it tells you when the batteries are set to wrap up. This will guide you from not allowing the robot stay out of juice all around. A battery screen guarantees that your aeronautical vehicle won't pass on in the most inauspicious spot.

•   **Mounting cushion**: Vibrations is completely reduced thereby improving the flight. This one is helpful particularly on the off chance that you are attempting to take pictures or recordings with your do-it-yourself drone.

- **Controller**: This gadget shares the force and orders the engines at every instance.

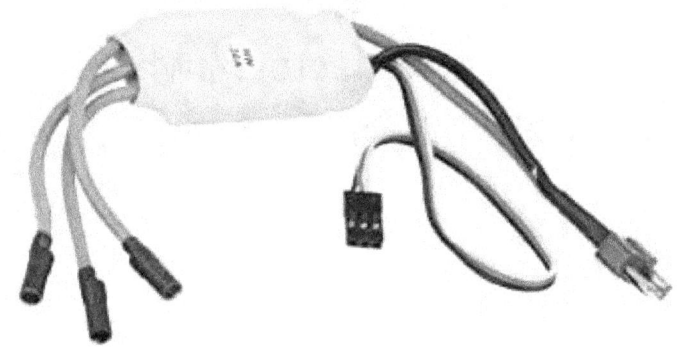

- **RC beneficiary**: In the event that you have a transmitter (which is ordinarily with you), you'll likewise have a recipient mounted on the robot.

- **Camera**: you will need a good camera if you want to take flying photographs and record the environmental factors while flying your robot. Recommended cameras are those that can take the quality 4K recordings.

According to importance, get the camera that suits your use. For excellent flying and getting good photography and video graph, you may require a gimbal for the camera.

• **USB key**: This is important to save the photographs and recordings.

Other useful parts that can be installed on your robot for better development are listed below;

I.   AWG silicone wires
II.   Battery charger
III.   Servo lead wire links
IV. Zip ties
V.   3M order strips
VI.   String locking mixes

Fabricating a robot comes in several ways and depending on the amount of genuine apparatus at hand.

The guide in this book will help you build your own quadcopter.

CHAPTER THREE

Robots comes in several types and shape, however many persons discover quadcopters to be more proficient, as they are not difficult to fly. Below is the step taken to fabricate your own quadcopter with other parts can be independently purchased'

STAGE 1: MAKING THE FRAME

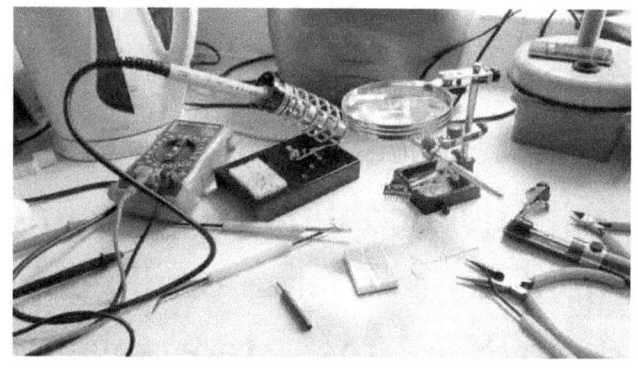

Every robot is expected to have casing, therefore our next line of action is to make a casing. For this reason, you can utilize various materials, for example, metal, plastic, or wood. These materials will help determine how strong you expect the robot to be.

On the off chance that you select wood for the casing, discover a wood board that is longer than 60 cm and around 25-30 mm thick. Cut up this board in such a manner to get two strips which are 60cm long and 30mm wide. These two lengths are needed to make the construction of your future quadcopter.

Intersection these two strips you'll make the X casing. Additionally, you will require a wooden sheet to make and add a rectangular piece in the focal piece of this edge. Its size ought to be 6×15 cm, and about 2mm thick.

Obviously, you can utilize different measurements on the off chance that you like, yet these will get you a quite pleasant quad. To interface these parts, you will require nails and paste. On the off chance that you choose to go with metal or plastic, the measurements are comparable yet the manner in which you interface the slats together will be extraordinary.

Use the below recommendations as a base for your venture:

- ✓ "Quadcopter Frame (Carbon Fiber) "
  - ✓ "Drone Frame (Carbon Fiber)"

- ✓ "FPV Frame that is made of Carbon Fiber"
- ✓ "Carbon Fiber Drone Frame"
- ✓ "Frame with Landing Gear made of Carbon Fiber"

STAGE 2: THE PROPELLERS, ELECTRONIC SPEED CONTROLLERS, AND MOTORS

The Electronic Speed Controllers), the engines, and the propellers are among the main components of a utilitarian robot. You are advised to visit approved dealers to get a guarantee, quality and dependability speed controller and motor. They should be the exact size of your robot, so uncovered this at the top of the priority list when getting them. Get help each time you are not too clear with any step during your installation.

When searching for the engines (or rotors), you should realize that multi-rotor drones produce more noteworthy speed and guarantee a steady flight, as every rotor works with other's pushed focuses. For instance, look at these rotors:

- ✓ "Emax RS2205 2600KV Brushless Motors"
- ✓ "Readytosky GT2205 2205 2300KV Brushless Motor"

- ✓ "HOBBYMATE 2204 QuadCopter Rotors Combo"

For the propellers, we recommend you purchase the metal 9-inch props you can discover at a truly reasonable cost available. These are tough and won't twist so effectively if the robot hits something during flight. Be that as it may, in the event that you need better execution, it is smarter to get carbon propellers. If you need great execution we prescribe you to get any of these:

- ✓ "Propellers made of carbon BTG Quick Release
- ✓ "Myshine 9450 Self-fixing Propeller Props"

I would recommend these extraordinary and stable Electronic Speed Controllers for your use as listed below:

✓ "A Airbot Omnibus F4 Nano - Flight Controller".

STAGE 3: GET THE MOTORS AND ASSEMBLE IT

The following thing you need to do is drill the openings in the edge for the engines, as per the distance between the screws openings on the engines. It is acceptable to make another opening that will permit the clasp and shaft of the engine to move unreservedly.

Be that as it may, you may avoid this activity if the engines previously accompanied mountings. Put the engine in the suitable spot and fix it to the casing utilizing the screws and a screwdriver.

STAGE 4: MOUNT THE ELECTRONIC SPEED CONTROLLERS

Subsequent to mounting the engines, you additionally need to mount the speed regulators. How might you do this? It is prescribed to interface the speed regulators on the base side of the edge because of a few reasons which include the usefulness of the robot. These reasons, among others, incorporate that it will "dump" the upper side of the robot where different segments ought to be added.

To fix the ESC to the casing, you need to utilize zip ties. Along these lines, your ESCs are secured and very much made sure about while flying.

STAGE 5: GET AND ADD THE LANDING GEAR

This stuff is a significant part when handling your UAV in light of the fact that it essentially diminishes the shock that may exist on the robot when it lands on a strong ground. It may be made in an unexpected way, yet you ought to be imaginative and make it in your own, remarkable way.

I suggest that you locate a metal line (around 6 creeps in breadth) and cut off (with the suitable devices) 4 rings that will be 1-2 cm in thickness. Obviously, the size of these rings ought to be as per the overall size of your robot. You would then be able to utilize conduit tape to fix these pieces to the edge.

In the event that you don't care for this metal line thought, you can likewise utilize different materials that are adaptable yet solid, for example, some new plastics, or anything that will decrease stun.

STAGE 6: THE FLIGHT CONTROLLER

Each flying robot should have a control framework. This electronic framework permits a robot to be steady noticeable all around while it gets wind and balancing.

There are two alternatives with regards to this progression:

To start with, and the simpler choice, is to purchase a prepared to-utilize regulator. With the subsequent choice being that you make it yourself.

For this work, you can utilize one of the accompanying source flight regulator projects listed below;

✓ Open Pilot CC3D: This heavenly open-source flight project is made to have 6- channels and the MPU-6000. It is exceptionally simple to set up and introduce, and there is a wizard that direct and drives you through the establishment.

✓ NAZE32: Very adaptable however somewhat stressful to set up. It has the high level fliers which improve the authority over your robot, yet you should ensure you can really set it up.

✓ KK2: This is quite possibly the most utilized tasks for this reason since it is less expensive than most different wellsprings of that kind. It accompanies LCD that depends on the high level AVR regulators. Hence, you can set it up without utilizing a PC. It also comprises of MPU6050 and a sensor, which permits you to compose your firmware. Nonetheless, KK2 needs manual tuning and it isn't helpful for RC fledglings.

On the off chance that you need to make a regulator yourself, you ought to select one of these tasks that best suits your requirements. Follow the connections above to do some more research, and look at the people highlights of each in more detail. It is extremely confounded to develop such a gadget and requires a specialist drone expert. However, in the event that you are capable, your robot will be simply a definitive "do-it" flying vehicle.

STAGE 7: YOU WILL NOW CHOOSE A CORRECT RC TX-RX (WIRELESS REMOTE CONTROL SYSTEM)

This is the controller framework that is expected to control a drone.

There are different accessible RC control frameworks in the market these days among which are;

1. Turnigy
2. Futaba
3. Spektrum
4. FlySky and lots more

Notwithstanding this framework, you'll likewise require a couple of channels for yaw, pitch, choke, and move, just as the extra channels on the off chance that you need to mount a camera control to your robot for some airborne photography.

STAGE 8: YOU WILL NOW MOUNT THE FLIGHT CONTROLLER

When you pick the specific flight regulator that is best for building your drone, you will now mount it. Mounting process is shown below;

you can put it on the highest point of the edge a specific way, yet you need to ensure that all the segments are fixed well prior to aligning your robot. For this reason, you can likewise utilize the zip ties which were referenced previously.

It is prescribed to put a little piece of wipe on the underside of the flight regulator since it retains and diminishes the vibrations from the engines. Subsequently, your robot will be steadier while flying, and strength is vital to fly a robot

Stage 9: YOU WILL NOW CONNECT THE OPEN PILOT TO YOUR DRONE

The following thing you need to do is to design and interface the flight regulator to the electronic speed regulators.

Likewise, you need to interface it to the controller. To perceive how to do this progression, you should locate a suitable instructional exercise video on the web for the specific flight regulator you have recently mounted.

Stage 10: CHECK OUT AND TEST YOUR DRONE

Before you at long last utilize your robot, you should be certain that everything functions admirably. Accordingly, you need to look at all the capacities before the primary flight. You can test the sensors just as different segments of your robot utilizing the uncommon Open Pilot GCS.

To ensure that everything functions admirably, you need to remove the props and make a little analysis with the controller. This guarantees that you can test the robot without taking a chance with the capability of breaking it.

For this test, you should locate an appropriate spot and attempt to move your robot inside its control distance. Focus on the zip ties and links to ensure that they are associated well. When all is well, your robot is prepared to fly!

Be certain not to compromise in this progression, it is basic to test everything in detail before really flying the robot. You would not need your robot's first trip to be its last all things considered!

Stage 11: TAKEOFF

This is the last venture. Prior to removing, the battery should be very much associated and all the segments should be fixed set up. For the dry run, you need to pick an area cautiously, since this airplane can cause genuine harms and can be harmed also. It is ideal to pick an open, level territory, so you don't risk harming anything with your robot, or the other way around. Likewise, you will guarantee that you can see your robot no matter the distance it moves to.

Spot your quad on the ground, put it into activity, takes the flight regulator, and begin with your first flight. It's suggested that you gradually choke up your robot, and fly it at low height for the absolute first time. Accordingly, on the off chance that it begins descending crazy, the harm won't be that critical.

In the event that the robot begins floating one way, you need to utilize the trims to make the fundamental flight adjustment. Likewise, you should evaluate diverse PID esteems to perceive how your robot functions in different contributions until you get precisely what you need.

STEP BY STEP INSTRUCTIONS TO FOLLOW IN HANDLING AND MAINTAINING YOUR DRONE

1. Pre-Flight Checklist

2. Have a reasonable Case/Backpack for your robot

3. Take consideration of your Batteries

4. Keep your product refreshed

5. Make sure your propellers are in acceptable condition

6. Keep your engines clean. ...

7. Fly your robot in great conditions

8. Regularly spotless your robot.

HOW TO MAKE A POWERFUL DRONE

1. Choose the right conditions to fly
2. The weight of drone must be reduced
3. Get a drone's propeller size that fit it
4. Get an upgraded battery battery
5. Charge the batteries a few hours remaining for the drone to flying
6. Battery care practices must be strictly adhere to

CHAPTER FOUR

CONCLUSION

This book will simply guide you through on how to build drone from scratch without preparation. You can research online for several ways of building drone for sale or for self benefits.

Making a drone and flying it yourself gives you a self confidence of doing thing yourself without assistance. Flying your personally constructed drone will encourage you to give it a better finish since you are not competing with anybody. It is expensive in building your own drone with a minimum cost of about $200 to $300 respectively. You will not regret getting this book for yourself development

THE END